DANGEROUS SEA CREATURES

Written by The Good and the Beautiful Team
Design by Robin Fight

BARRACUDA

THE BARRACUDA DOESN'T USUALLY ATTACK HUMANS, but it is very dangerous to fish! Unlike many other ocean predators which hunt by smell, the barracuda hunts by sight. This fish has teeth—two rows, actually—well situated in its large mouth with an extended lower jaw. The barracuda is also very fast, reaching burst speeds of up to 56 km (35 mi) per hour.

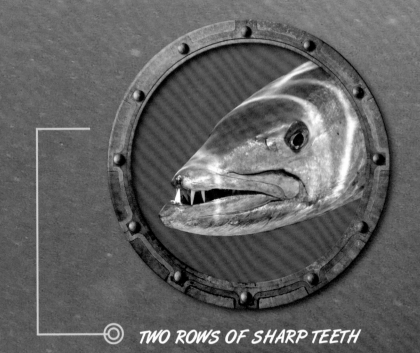

 Although not generally eaten, the barracuda is often fished for sport due to its feisty nature.

TWO ROWS OF SHARP TEETH

 Attacks are not common, but the barracuda can be dangerous to humans. This is because a barracuda could mistakenly attack shiny objects on a diver or even compete for a fish on a spear.

SPEEDS UP TO 56 KM (35 MI) PER HOUR

 The barracuda lives worldwide near the shore and in the open ocean.

 A young barracuda can change color to blend in with its surroundings.

BLUE-RINGED OCTOPUS

THE BLUE-RINGED OCTOPUS IS A TINY TERROR found in shallow depths of the Indian and Pacific Oceans. There are four types of this invertebrate, each measuring only 12–25 cm (5–10 in) in length, including its arms. When frightened or hunting prey, it can either inject or release a cloud of powerful venom which paralyzes the victim. The faint blue rings marking its body will flash bright blue to warn predators and larger creatures when it feels threatened.

 Like other octopuses, it has three hearts, and its blood is transparent blue.

 The tetrodotoxin (the same toxin found in puffer fish) it releases is produced by bacteria in its salivary glands and is more powerful than any land animal's toxin.

 The blue-ringed octopus is shy and will often hide in marine crevices or shells.

MEASURES ONLY 12–25 CM (5–10 IN) IN LENGTH

POWERFUL VENOM PARALYZES PREY

RINGS FLASH BRIGHT BLUE WHEN THREATENED

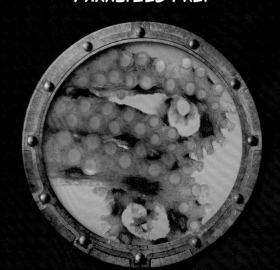

BOX JELLYFISH

ABLE TO MOVE ON ITS OWN INSTEAD OF DRIFTING

WHEN YOU THINK OF THE MOST VENOMOUS CREATURES IN THE WORLD, you may imagine a rattlesnake or a spider, but few snakes and spiders are as deadly as the box jellyfish. It uses its highly toxic venom to paralyze or kill fish and shrimp for its dinner. The box jellyfish is much more advanced than most jellyfish; God gave it the ability to move on its own rather than just drifting, and unlike other jellyfish, it has eyes!

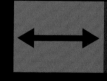 Living for up to a year, box jellyfish can grow up to 3 m (10 ft) long.

 On average 100 human deaths per year are caused by the box jellyfish.

 The box jellyfish does have a predator—the sea turtle, which is not hurt by its venom.

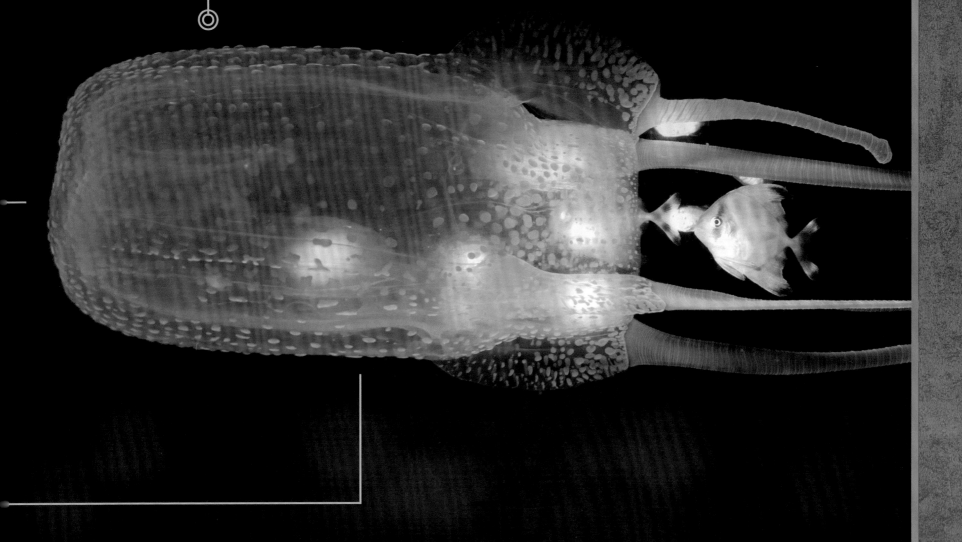

 Called the "bell," the box jellyfish's head has four groups of six eyes. That's a total of 24 eyes!

CROWN-OF-THORNS STARFISH

PREYS ON CORAL POLYPS, DIGESTING NUTRIENTS THROUGH ITS STOMACH

WHEN YOU THINK OF DANGEROUS SEA CREATURES, you probably think of those dangerous to humans. While the crown-of-thorns starfish can certainly harm humans with its toxin-filled spines, the real danger caused by this invertebrate is to the stony coral species. The crown-of-thorns starfish preys on stony coral, digesting the nutrients through its stomach and leaving only a coral skeleton. The starfish can turn its stomach inside out through its mouth to eat coral polyps found in reefs of the Pacific and Indian Oceans.

 Due to its venomous spines, this sea star has few natural predators. The predators it does have are quickly disappearing due to overfishing.

 An adult of this species can have anywhere from 12 to 23 arms.

The crown-of-thorns starfish is the second-largest starfish in the world; on average it is 45 cm (18 in) across.

The crown-of-thorns starfish gets its name from the hundreds of venomous spines on its body, which resemble the biblical crown of thorns placed on Jesus before His crucifixion.

TOXIN-FILLED SPINES CAN HARM HUMANS AND PREY

SEA SNAKE

CUTTING ITS PADDLE-LIKE TAIL BACK AND FORTH

to move swiftly through the warm coastal waters of the Indian or Pacific oceans, the sea snake, or coral reef snake, may have the most potent venom of all snakes found in or out of the water. Growing to around 1–1.5 m (3–5 ft) long, the sea snake is usually found in shallow waters, such as those of coral reefs, and is entirely aquatic. It does not have gills, however, and must come to the surface to breathe.

 The body of the sea snake is "flattened" and smooth in appearance, as its ventral scales are smaller which makes it challenging to move outside of the water.

 The only genus that has kept the larger ventral scales is the sea krait (genus: *Laticauda*), allowing it to spend much of its life cycle on land, including laying its eggs there.

 Most sea snakes can breathe through the top of their skin. The black and yellow sea snake can obtain roughly 25% of its oxygen this way, allowing this snake to swim in deeper water and for longer lengths of time without surfacing for air.

MOSTLY FOUND IN SHALLOW WATERS

MUST COME TO THE SURFACE TO BREATHE

GREAT WHITE SHARK

WITH A MOUTHFUL OF 300 SHARP TEETH that are serrated like a steak knife to rip up its food, the great white shark is the ocean's most fearsome predatory fish. Although it has a reputation of being the most likely shark to attack humans, we are not its choice prey. Often, it bites humans out of sheer curiosity. It would much rather snack on a tasty seal, dolphin, sea turtle, or even a small whale. This shark is extremely strong—especially its tail, which can propel it through the water at up to 24 km (15 mi) per hour!

The largest great white shark on record is a 6 m (20 ft) long, 2,300 kg (about 2.5 ton) female nicknamed Deep Blue.

A great white shark can cycle through 20,000 teeth in its lifetime.

A great white shark can detect one drop of blood in 95 L (25 gal) of water from up to 5 km (3 mi) away.

SHARP, SERRATED TEETH RIP OPEN PREY

EXTREMELY POWERFUL TAIL PROPELS ITS MOVEMENT

SPEEDS UP TO 24 KM (15 MI) PER HOUR

PORTUGUESE MAN-OF-WAR

IS ACTUALLY A GROUP OF MANY TINY POLYPS FUNCTIONING AS A WHOLE

DID YOU KNOW THAT THE PORTUGUESE MAN-OF-WAR ISN'T REALLY A JELLYFISH? While it looks like a single animal, this wonder of nature is actually a group of many tiny animals called *polyps*. You can think of these polyps as clones that each have specific jobs. Some of them make up the man-of-war's tentacles, some digest food, and one serves as its float. Although the man-of-war's venomous tentacles are extremely painful when touched, its sting rarely kills humans.

⟷ The tentacles of this intimidating creature are usually about 9 m (29.5 ft) long, but they can reach lengths of 50 m (164 ft). That's half the length of a football field!

 Groups of these creatures are called legions and can contain as many as 1,000 man-of-wars.

Filled with gas and used as a sail, the Portuguese man-of-war's float is colorful and useful.

VENOMOUS TENTACLES CAUSE EXTREME PAIN WHEN TOUCHED

PUFFER FISH

INGESTS AIR OR WATER
TO INFLATE ITS STOMACH,
CHANGING ITS SHAPE
TO THAT OF A LARGE BALL

RANGING IN SIZE FROM 2.5 CM (1 IN) TO OVER 60 CM (2 FT) in length, this clumsy swimmer does not rely on speed to elude predators. Instead, the puffer fish ingests large amounts of water or air to inflate its stomach, transforming itself into an inedible ball several times its original size. Most predators unfortunate enough to succeed in eating a puffer fish will soon die as the puffer fish contains enough poisonous tetrodotoxin in its body to kill 30 adult men.

 Although some puffer fish can live in fresh water or brackish water, most of the over 120 species of puffer fish live in tropical and temperate oceans.

 Newly discovered, the white-spotted puffer fish creates fascinating circular designs on the seafloor.

The puffer fish survives on a diet of primarily invertebrates and algae, but it can break open and eat shellfish with its beak-like teeth.

RED LIONFISH

FANNED-OUT
FINS

RELAXED
FINS

RESEMBLING A RED-AND-WHITE-STRIPED LION'S MANE, large featherlike fins stick out from the red lionfish's body. Some of those fins contain venomous spines that the red lionfish can use to sting an attacker. The red lionfish can also fan out its fins, distracting an invertebrate prey and forcing the prey into a tight space. The red lionfish then blows directed jets of water at the invertebrate before consuming it whole.

 The mature red lionfish is between 30 and 38 cm (12 and 15 in) long, weighs up to 1.2 kg (2.6 lbs), and lives up to 15 years.

 Native to the Indian and Pacific Oceans, the red lionfish was introduced to the American Atlantic coasts by humans.

 The red lionfish can be eaten, but the primary reason for capturing it is for displaying in aquariums.

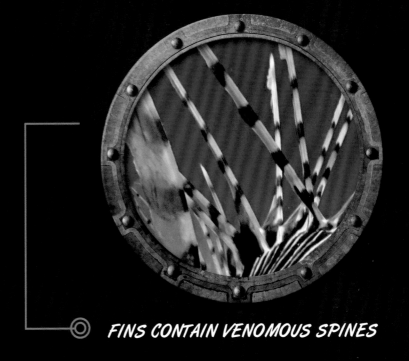

FINS CONTAIN VENOMOUS SPINES

FANS OUT ITS FINS TO DISTRACT PREY

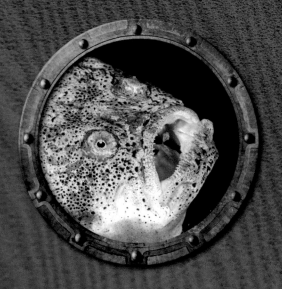

STARGAZER

DISGUISES ITSELF BY HIDING IN THE SAND ON THE SEAFLOOR

LARGE, UPTURNED MOUTH VACUUMS IN PREY

DON'T LET THE LOVELY NAME FOOL YOU — the stargazer is a fierce predator! This scary fish hangs out in the sand on the seafloor, using a wormlike piece of flesh inside its upturned mouth to lure an unsuspecting fish or crab into swimming closer. Then it opens its mouth quickly enough to vacuum the prey into its mouth and eats it whole. Some species have venomous spines on their backs, which they use to inject venom into their prey, while others have an organ that can deliver a 50-volt electric shock.

 The stargazer is often called "the meanest thing in creation."

 Since its venom is not poisonous when ingested, some cultures consider the stargazer a delicacy.

 The scientific name includes the Latin word *astroscopus*, which means "one who aims at the stars."

 Though most fish breathe through their gills, the stargazer breathes through its nostrils.

What appear to be terrible, pointy teeth are actually just pieces of fringe to keep sand out of its mouth. Its actual teeth are tiny because it doesn't need to chew its food.

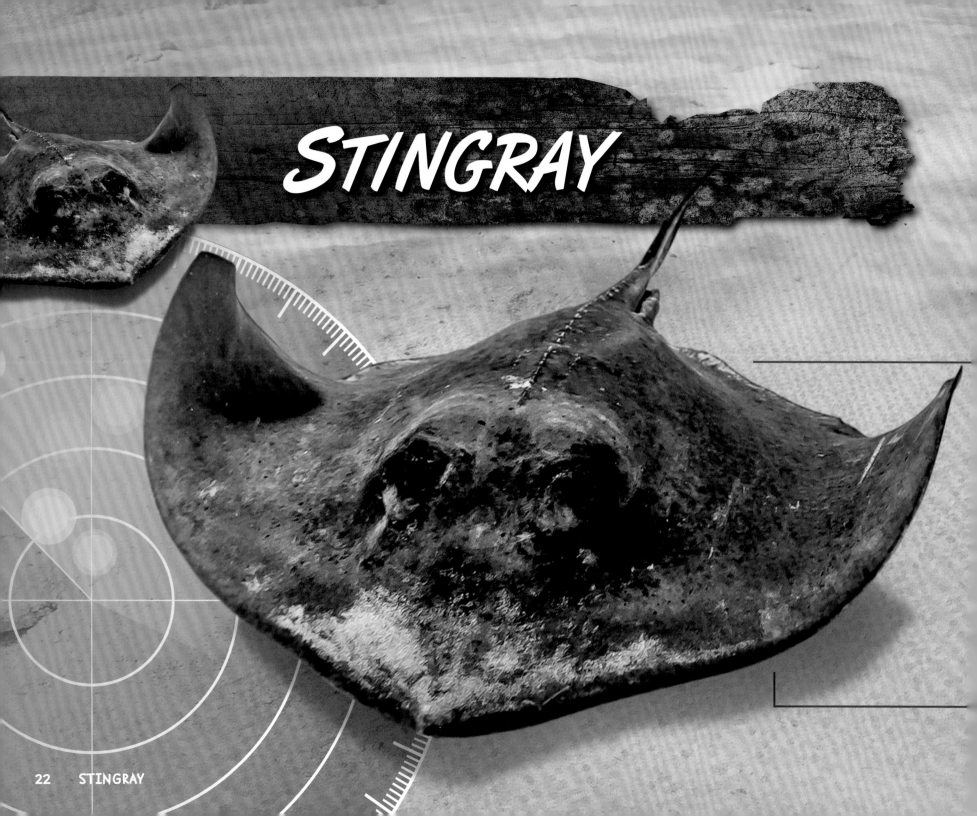

STINGRAY

THE BEAUTIFUL STINGRAY IS A FLAT, DISK-SHAPED FISH with unique fins resembling wings. The stingray belongs to the group called elasmobranchs. Because the stingray is made only of flexible cartilage, it glides through the ocean like a gentle wave in warm, tropical waters. While the stingray spends a quiet life as a bottom dweller, often partially buried in the sand, its whiplike tail contains barbed, venom-filled spines, which can inject unsuspecting swimmers and pack a painful sting.

 A stingray's coloration reflects the seafloor, allowing it to stay camouflaged from its many predators, including sharks and seals.

 The stingray eats small crustaceans, snails, clams, fish, shrimp, and other creatures.

 This amazing creature ranges in type and size from the smallest short-nosed electric ray weighing 400 g (14 oz) to the oceanic manta ray weighing 2,000 kg (4,409 lbs).

HIDES IN THE SAND ON THE SEAFLOOR

WHIP-LIKE TAIL WITH BARBED, VENOM-FILLED SPINES

GLIDES EASILY DUE TO ITS BODY OF FLEXIBLE CARTILAGE

STONEFISH

EASILY CAMOUFLAGED AMONG ROCKS AND STONES

THE MOST LETHAL FISH IN THE WORLD

IN THE SHALLOW WATERS OF THE INDO-PACIFIC, quietly buried in the sand, lies the most lethal fish in the world. It awaits its prey, well camouflaged, with its warty skin and mottled coloring concealing it among the rocks. Watch out! Many have unwittingly stepped on this stonelike fish, triggering the needle-sharp spines along the stonefish's back to inject deadly venom into its victim. This venom is so toxic that, without treatment, the person could die within two hours!

 The stonefish gets its name from its appearance because it looks just like a rough and pitted stone.

 Thirteen dorsal spines line the stonefish's back, a venom sac sitting at the base of each sharp spike. With the slightest amount of pressure, the venom is injected into the stonefish's prey.

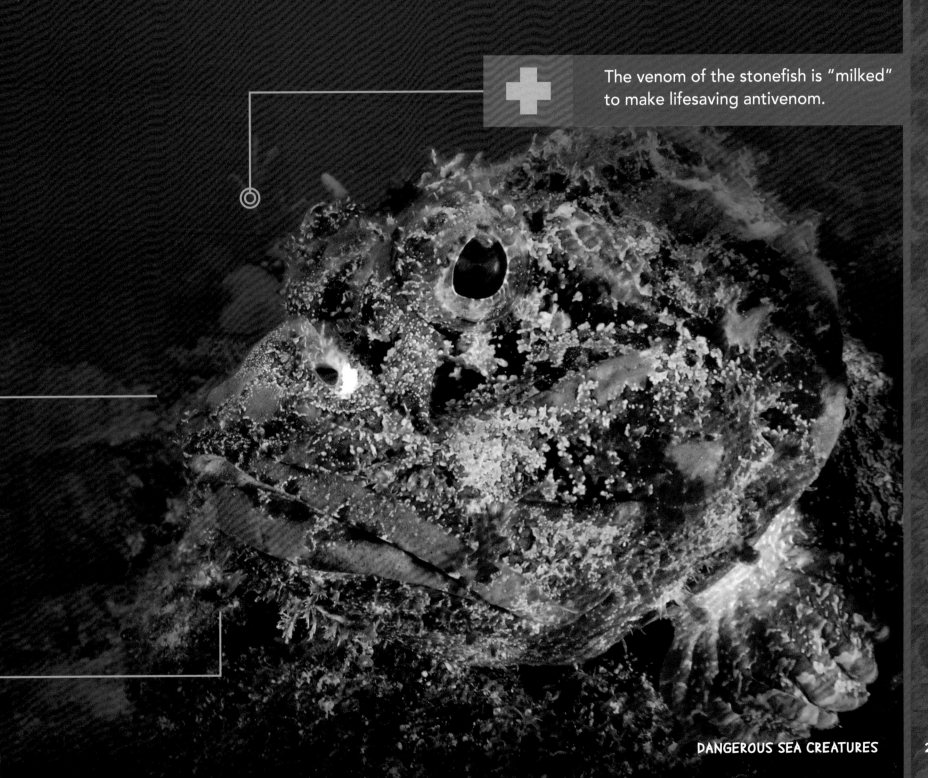

The venom of the stonefish is "milked" to make lifesaving antivenom.

STRIPED SURGEONFISH

THE STRIPED SURGEONFISH gets its name because of a retractable spine within its tail that is as sharp as a surgeon's scalpel. Ouch! This spine helps the surgeonfish defend itself from predators. This colorful fish lives along reefs and is easily recognized by the yellow and black stripes on its upper body and pale blue to purple coloration on the lower quarter of its body. Can you imagine watching this beautiful fish as it darts around the Great Barrier Reef searching for food?

 The striped surgeonfish will grow to be 38 cm (15 in) long.

 These fish can be found in the Indian and Pacific Oceans.

 While most surgeonfish are not venomous, some species are.

 Other names for the striped surgeonfish are clown surgeonfish or clown tang.

SHARP, RETRACTABLE SPINE IN TAIL

YELLOW-AND-BLACK-STRIPED MARKINGS

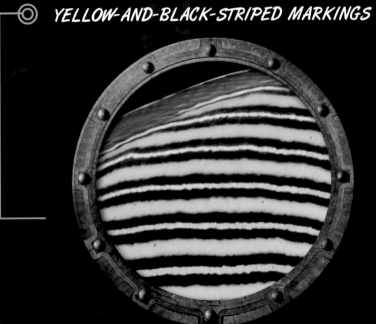

TIGER SHARK

A tiger shark is so named because of the distinctive striped pattern found on juveniles of the species. However, the markings fade as the shark gets older.

WHAT OCEAN PREDATOR has been known to eat sea turtles, seals, ocean birds, dolphins, rays, sharks, and even garbage? The tiger shark! Because the tiger shark isn't very discerning in its diet and generally stays near coastal areas, it is considered the second most dangerous shark to humans. Only the great white shark has been guilty of attacking humans more often than the tiger shark. The tiger shark has sharp, jagged teeth, and its jaws are strong enough to crack a sea turtle's shell!

Some of the largest tiger sharks ever recorded were 5–8 m (16–26 ft) long and weighed around 907 kg (2,000 lbs).

In the wild this shark can live 15 years or more.

The tiger shark is a scavenger, and it has been found with old tires, burlap sacks, and even license plates in its stomach.

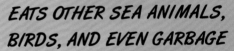
EATS OTHER SEA ANIMALS, BIRDS, AND EVEN GARBAGE

SECOND-MOST DANGEROUS SHARK TO HUMANS

JAWS ARE STRONG ENOUGH TO CRACK A SEA TURTLE SHELL

YELLOW SEA ANEMONE

ATTACHES ITSELF TO HARD SURFACES, SAND, OR OTHER ANIMALS

THE YELLOW SEA ANEMONE LOOKS LIKE A BEAUTIFUL OCEAN FLOWER, but don't let its beauty fool you! This invertebrate sea creature has venomous tentacles that paralyze its prey and create discomfort for larger, more threatening creatures. Related to corals and jellyfish, the anemone is found around the world in shallow tide pools or as deep as 10 km (6.2 mi) below sea level. It generally attaches itself with its base, or foot, to a hard surface, sand, or even another creature.

 Some sea anemone species can live more than 50 years!

 Certain marine creatures, such as clownfish, are covered with a mucus that allows them to live in and around the anemone without being harmed.

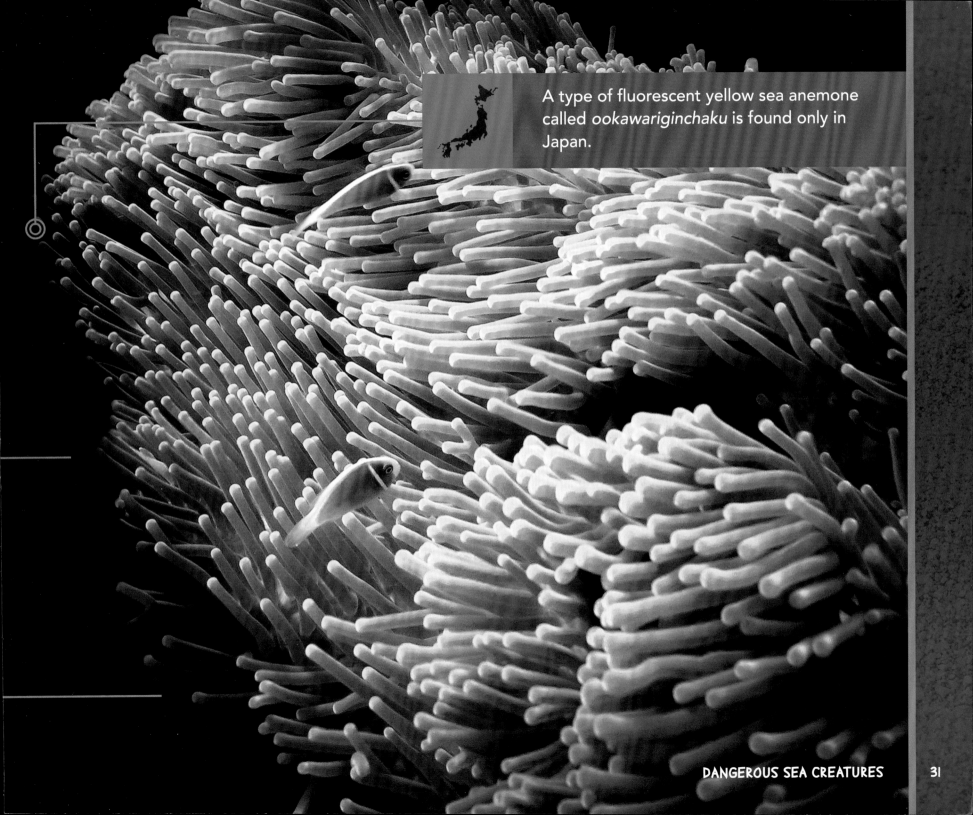

A type of fluorescent yellow sea anemone called *ookawariginchaku* is found only in Japan.